REAL SPACESHIPS!

How Spaceships Work

Contents

WHAT IS A SPACESHIP ?

A spaceship is a vehicle that is made to fly in space. Spaceships are manned by a crew of astronauts

Figure 1 Different types of spacecrafts

USE OF A SPACESHIP ?

Spaceships are designed to carry people and cargo. They can explore new places like the moon and take supplies to the International Space Station.

Figure 2 Astronauts in Space

But have you ever wondered how spaceships work and what they can do?

Get ready to learn everything you ever wanted to know about space-ships

THE SPACE AGE

The history of spaceships begins with the Space Age in the 1950s. During this time, the United States and the Soviet Union competed to see who could get to space first.

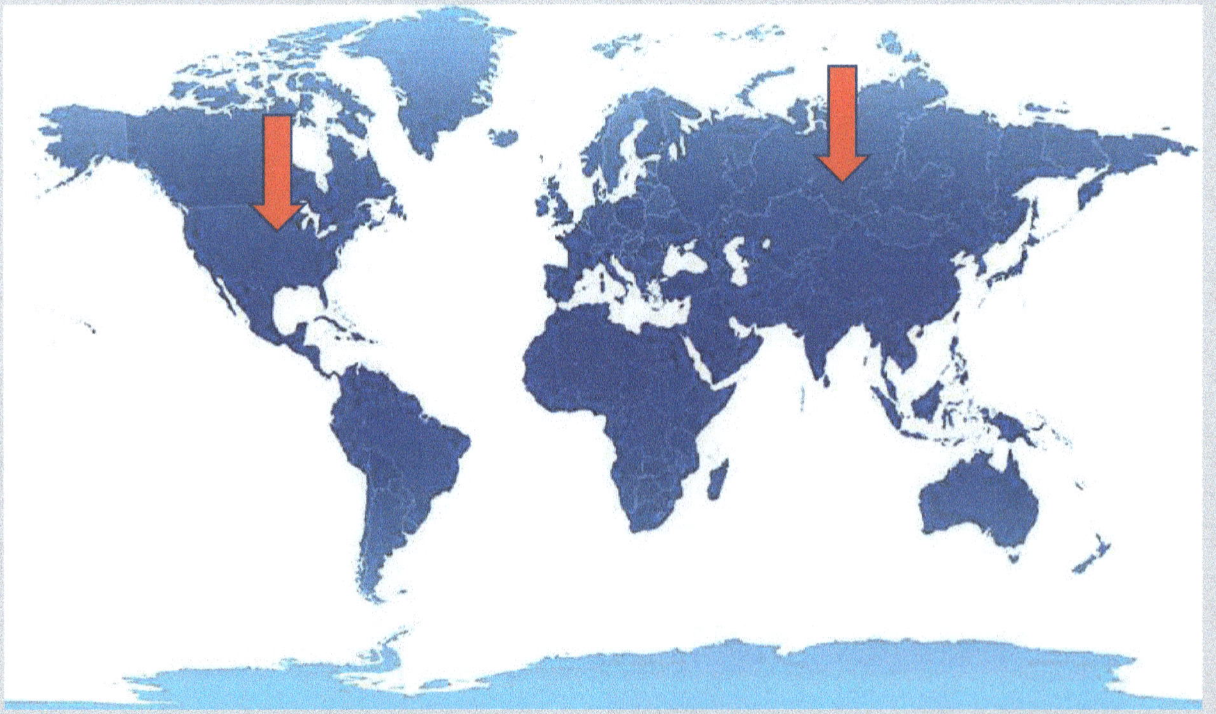

Figure 3. U.S. and Soviet Union were the first to send humans into space

Definition: A satellite is an unmanned object that is launched into space. It then orbits the Earth and can be used to gather and send information.

The Soviet Union made the first move. They launched the first satellite into space in **1957**. It was called **Sputnik I**.

Figure 4. Sputnik I in orbit

In **1958**, the U.S. created NASA (National Aeronautic Space Act). Its goal was to send a manned spacecraft into space.

Figure 5. NASA Headquarters

A year later, America launched the **Mercury program** and the Soviets launched the **Vostok program**. Both started creating spacecraft.

Figure 6. Vostok Program by Soviet Union

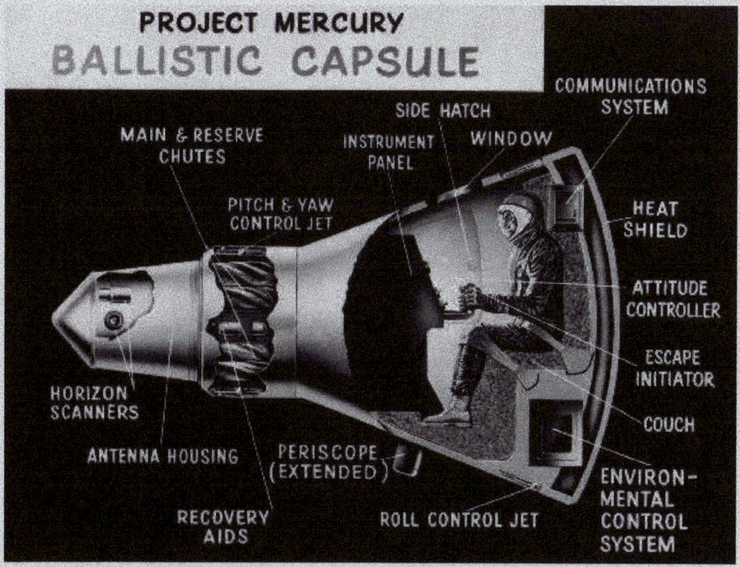

Figure 7. Mercury Program by the U.S

Russian **Yuri Gagarin** was the first man in space in **1961**. He beat American Alan Shepard by only a few weeks.

Figure 8. Yuri Gagarin. First man in space.

MAN ON THE MOON

For the first several years of the Space Age, the Soviets seemed to be winning. They launched the first satellite and the first animals, man, and woman into space. But America would catch up by the end of the 1960s.

In 1969, America won the Space Race when they finally landed a spaceship on the moon. Astronauts **Neil Armstrong** and **Buzz Aldrin** became the first men to walk on the moon.

As **Neil Armstrong** said, *"That's one small step for man. One giant leap for man-kind."*

Figure 9 Neil Armstrong on the moon

FROM THE MOON UNTIL NOW

In the 1980s, NASA launched the space shuttle program. The reusable space shuttles would make 24 trips into space over the course of six years. The program was retired in 2011

Figure 10. The Space Shuttle Challenger

Since 2011, there hasn't been as much focus on going to the moon. Instead, NASA and other organizations are focused on exploring other parts of our solar system and even beyond.

Figure 11 Artist rendering of space

SHIPS MADE IN THE US

Engineers all over the country are designing new types of spaceships. Who knows where these new ships may take us or how they may change our lives?

ORION

Orion is NASA's newest spacecraft. It's a manned ship designed for deep-space travel with plans to travel beyond the moon and maybe to Mars.

Figure 1 2 Orion spaceship

DRAGON

The Dragon is a ship developed by SpaceX and it comes in both Crew and Cargo models. It recently made a trip to the ISS with two astronauts. The Dragon can carry up to seven crew members and lands in the ocean when returning to Earth.

Figure 13 Dragon Spaceship

STARLINER

Boeing (known for its jets) designed the Starliner ship. It's a capsule built in conjunction with NASA. It can hold up to seven passengers and currently takes crew and cargo to the ISS

Figure 14 Starliner.

SpaceShipTwo

SpaceShipTwo is intended to be the first ship in space tourism. It's set to launch in 2022 and will take tourists into suborbital space

Suborbital space: Suborbital spaceships go to outer space, but the trajectory crosses over with the Earth's atmosphere. They never reach escape velocity.

Figure 15. Spaceship Two

NEW SHEPARD

New Shepard is a vertical launch and land spaceship designed for space tourism. It's created by the company Blue Origin, led by Amazon founder, Jeff Bezos. It currently goes into suborbital space, but tickets aren't publicly available.

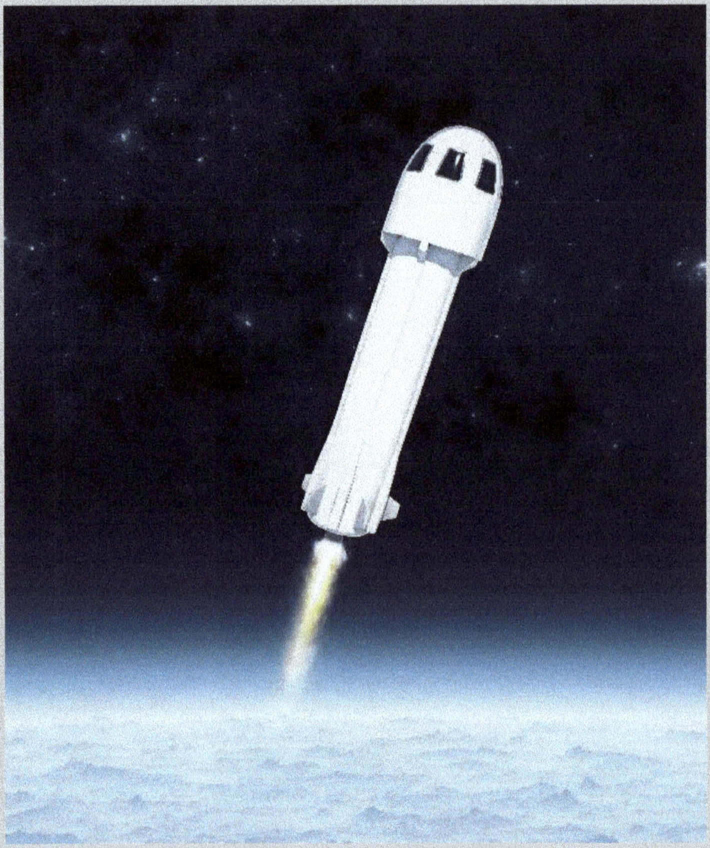

Figure 16. New Shephard

DREAM CHASER

Dream Chaser recently released its first ship, a non-capsule private orbital spacecraft called **Tenacity**. It launches from a rocket but lands like an airplane. It'll carry cargo to and from the ISS and is set to launch in 2022.

Figure 17. Dream Chaser Spaceship

PARTS OF A SPACESHIP

Spaceships require a lot of parts to get them into space! Let's look at some of the most important parts.

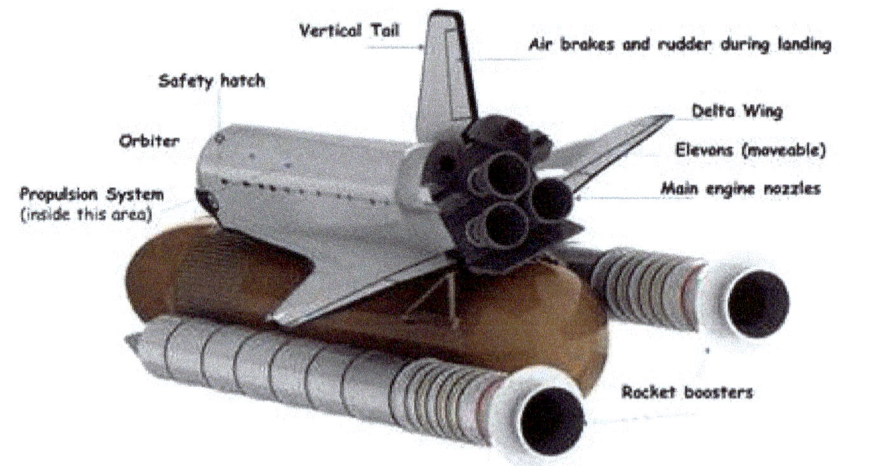

Figure 18a. Parts of a spaceship

Propulsion System

The propulsion system is what launches the spaceship into space. Most spaceships currently use chemical rocket fuel. It's heated and then moves out the back of the ship, propelling it upward.

Orbiter

The orbiter contains part of the spaceship where the crew stays and other essential equipment. It can go into orbit and then return to Earth, landing like a glider. A pressurized cabin keeps the crew safe - they work, control the ship, sleep, eat, and relax in this area.

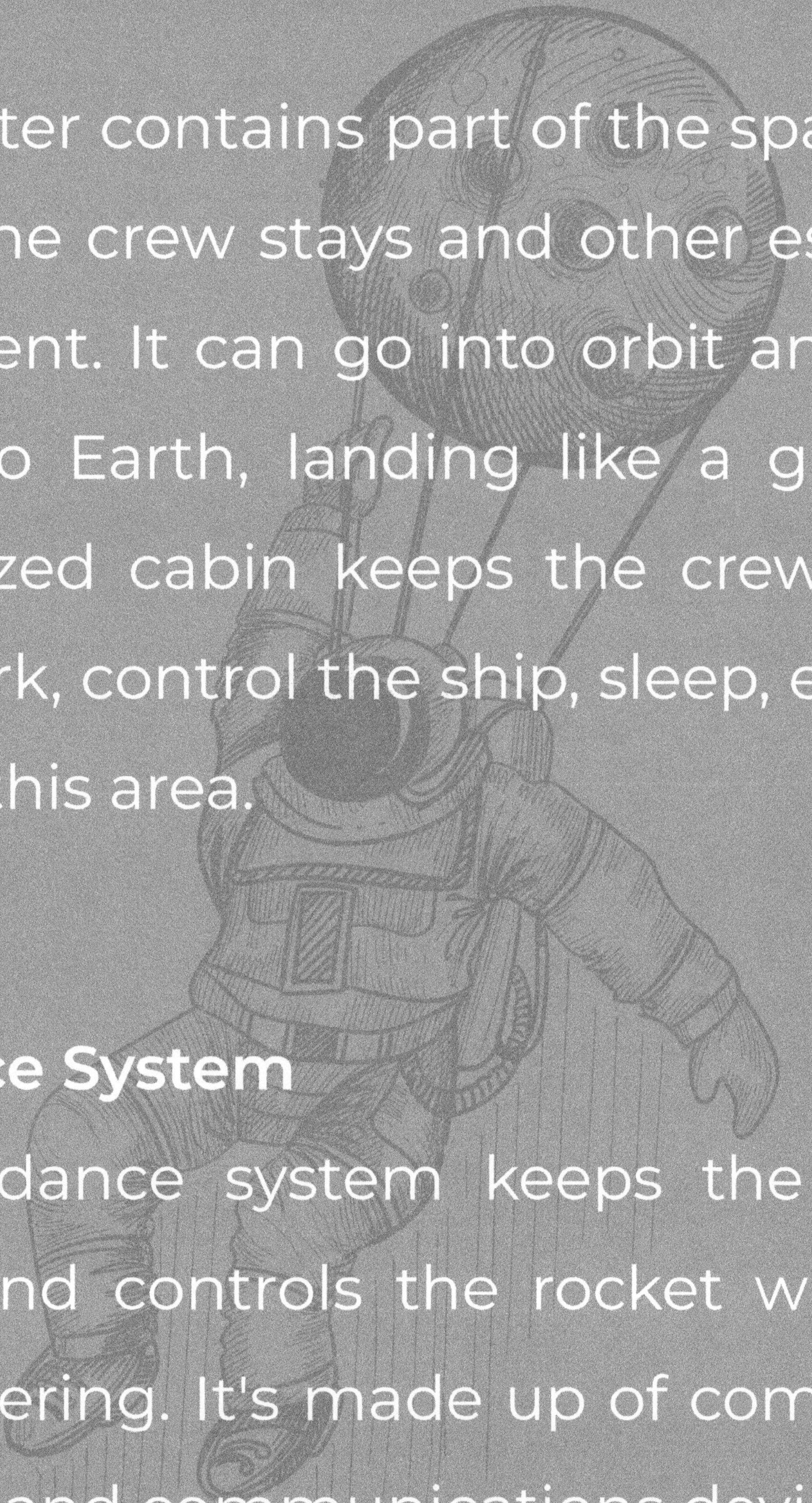

Guidance System

The guidance system keeps the rocket stable and controls the rocket while it's maneuvering. It's made up of computers, sensors, and communications devices.

Fuel Tank

You can think of the fuel tank, also called the external tank, as the gas tank of the orbiter. It supports the ship during launch and is the only part that isn't reused. The ship jettisons the tank several minutes after launch

Entry Hatch

The entry hatch is a small door for the crew to enter and exit the ship.

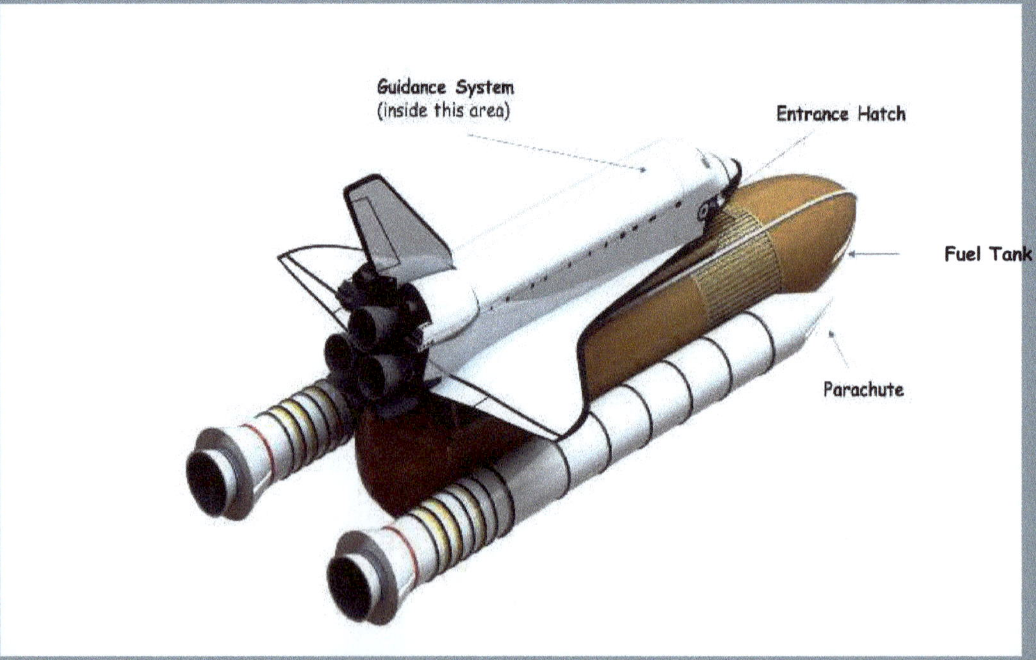

Figure 18b. Parts of a spaceship

Main Parachute

The main parachute is part of a set of parachutes designed to slow the ship down when it's returning to Earth. Ships can be going at 25,000 miles per hour or more upon re-entry. The parachutes can slow them down to 17 miles per hour.

Instrument Panel

The instrument panel contains all of the controls that the crew may need onboard. They can monitor all of the systems to make sure everything is going correctly, initiate launch sequences, and even put the ship into manual.

Pilot's Hand Controls

The hand controls allow the crew to control the ship. They usually need to do some adjustments during the ascent and re-entry.

Figure 19. Onboard Instrument Panel and hand controls

Porthole

Portholes are small round windows that allow the crew to see outside the ship. Spaceship portholes have to be made out of a special glass because they experience extreme temperature changes. They can be opened, but no one has ever tried.

Figure 20. Porthole

Escape Rocket

Things can go wrong during a spaceship launch. The escape rocket is designed to let the crew escape in the event of an emergency. It's connected to the capsule.

Capsule

The capsule is the small part of a spaceship where the crew stays. It's part of the orbiter

Figure 20. Capsule

WHAT ARE SPACESHIPS MADE OF?

Lots of materials go into making a spaceship. These materials have to be durable enough to withstand extremes in pressure and temperature while also being lightweight.

Composite Panels

When spaceships re-enter the Earth's atmosphere, they experience extreme heat. Composite panels were designed to withstand this heat. They're made of ceramic with a carbon bottom to keep them from getting too hot.

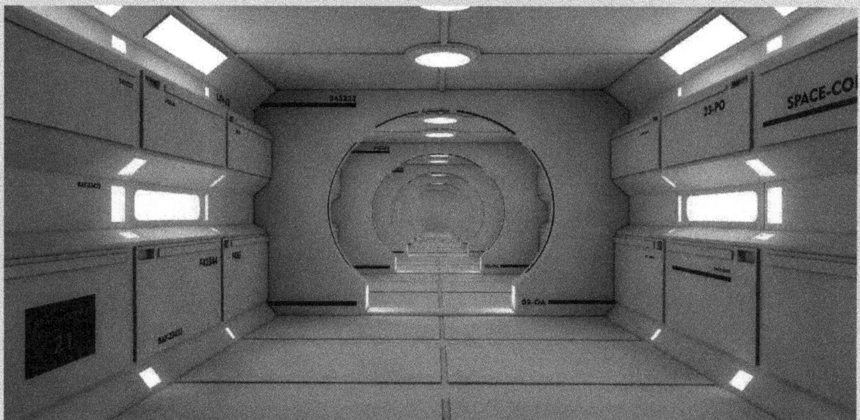

Figure 21. Interior of a spaceship

Self-Healing Panels

While composite panels work well, they do tend to form small cracks over time. Scientists are working on a composite material that will heal itself if it cracks.

Figure 22. Self-healing panels used on automobiles

Vanadium

Vanadium is often added to titanium to make an alloy. It creates one of the lightest alloys that is still incredibly strong.

Aluminum

Aluminum is a common metal used in spaceship construction. It's among the lightest and most durable metals.

Figure 23. A Titanium Panel

Titanium

Titanium is light (though not as light as aluminum) and incredibly strong. It's strong enough to attach heavy loads and is resistant to extreme temperatures.

HOW SPACESHIPS WORK?

Now you know all about the parts of a spaceship and the materials used to make them - but how exactly do they work?

Figure 24. Space shuttle Lift off.

Liftoff

Getting a rocket to lift off is a challenge. The normal methods we use for flying - like pushing against the air and using combustion engines - don't work in the vacuum of space. The current solution is the rocket.

Rockets work by using chemicals that release exhaust fumes when they explode. The rocket has a combustion chamber where the fuel is kept. The fuel, also known as a propellant, can be a solid, liquid, or gas. Different propellants are used for different ships and situations.

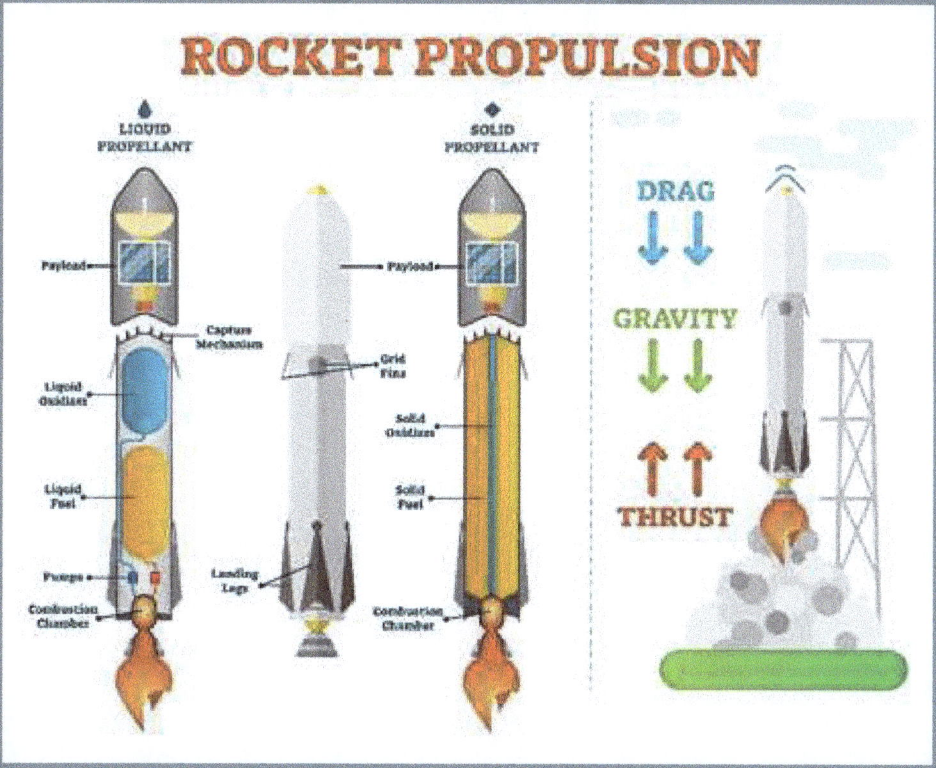

Figure 25. Rocket Propulsion Diagram

An oxidizer is added to the combustion chamber and when mixed with the fuel, it explodes. This explosion creates exhaust, or gases made by the reaction.

The exhaust is pushed out the bottom of the rocket and the rocket is pushed up. This force is called thrust and it can be explained by Newton's Third Law of Motion. Newton's third law states that for every action, there is an equal and opposite reaction. When the exhaust moves down, the spaceship moves up.

WHAT IS A SPACE MISSION?

There are four main stages to a space mission, although each mission has separate goals and may do different things while in space.

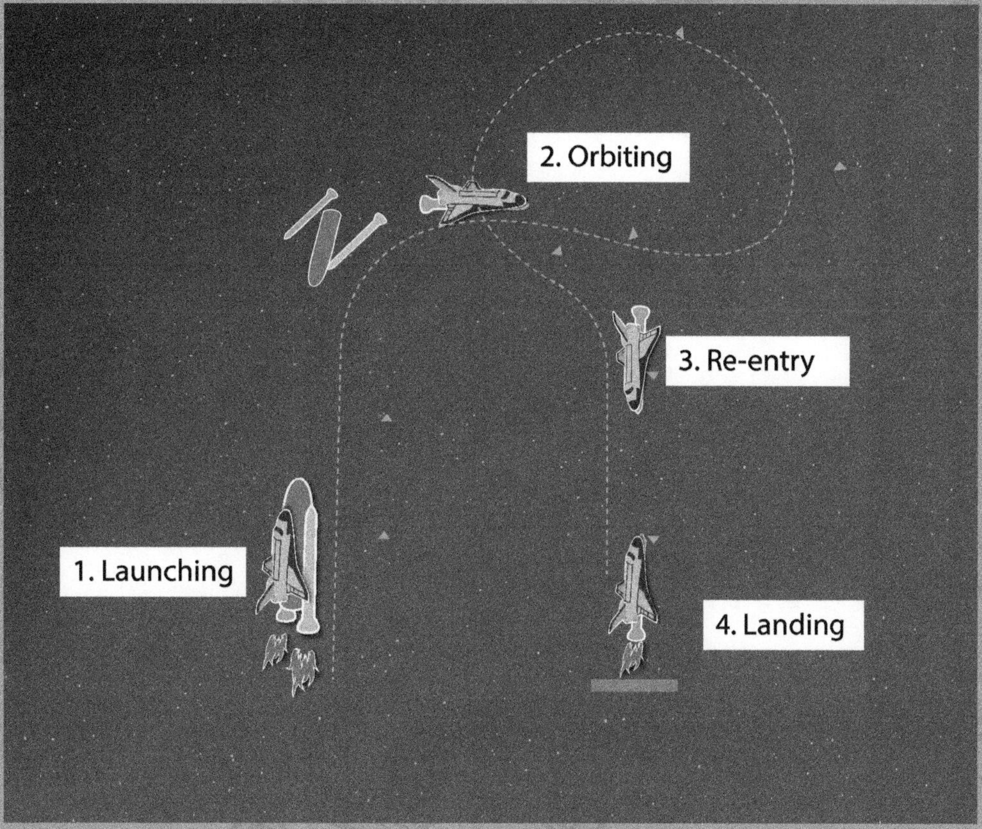

Figure 26. Diagram of a space mission

1. Launching

The spaceship is launched out of Earth's atmosphere and into space.

2. Orbiting

Once the spaceship is in space, the engines shut off. It doesn't need them because there's no air in space to cause friction. It follows a trajectory. However, its orbit continuously changes the trajectory. A team on the ground works to figure out where the ship is and predict where it will go.

3. Re-entry

Re-entering Earth's atmosphere is just as complicated as exiting. If the spaceship is going too fast, it'll burn like a meteorite. To slow down, the ship turns around and flies backwards to get out of its orbit.

The belly of the ship goes down first, which creates drag to slow it even more. The friction created by drag exposes the ship to 3000 degrees. Fortunately, it has insulation panels to protect it.

4. Landing

Once back in the atmosphere, the ship releases parachutes to slow it for a safe landing. Some ships have wings that extend to allow them to land like an airplane. Others land in the ocean.

WHY HUMANS WANT TO GO TO SPACE

There are many reasons humans want to go into space. We're curious creatures and we're always wanting to learn more. We've explored only a tiny amount of space, so we want to find out what's out there.

Figure 27. View of space

Just as the explorers set sail to find new lands, astronauts also want to explore new lands. We don't know what is out there, so we want to see what may be on other planets and moons.

On Earth, we have many resources we use that are nonrenewable. That means that when they're gone, they're gone. Space may provide more resources for us as we run out. We may even want to set up colonies on other planets in case Earth becomes inhabitable.

Figure 28. Illustration of Astronauts on Mars

WHERE DO SPACESHIPS GO?

Currently, spaceships have gone either to the **moon** or the **ISS (International Space Station).**

The Moon:

In the US, only 24 astronauts have been to the moon. The last moon trip was in 1972. Why has it been so long since we've gone to the moon? Mostly because of money. It's expensive to send humans to the moon, and there isn't a lot of money in the budget. However, there are private companies that are planning moon trips in the near future

Figure 29. The Moon

Figure 30. The International Space Station (ISS)

The ISS :

Spaceships often go to the ISS (International Space Station), taking both crew and cargo. The ISS was built by several different countries with the purposes of conducting experiments in space. These experiments will help us advance in our goal to travel deeper into space.

SPACESHIPS: FUN FACTS

1 The fastest spaceship is the Parker Solar Probe. It has reached speeds of over 330,000 miles per hour as it moves toward the sun. It's predicted to reach over 450,000 mph.

Figure 31. Parker Solar Probe

2 Voyager 1, a space satellite, is currently over 14 billion miles away and still going. It has gone farther than any other spaceship. It left the solar system in 2012.

Figure 32. Voyager 1 Launched in 1977

3 Our closest stellar neighbor is Alpha Centauri. It's 4.25 light years away, meaning that if humans are going to leave the solar system, we'll have to travel at the speed of light. But is this possible?

Figure 33. Alpha Centauri in the night sky.

HOW FAST DOES LIGHT TRAVELS?

Light travels at **186,282 miles per second**. While we have traveled faster than the speed of sound, we will most likely never be able to travel as fast as or faster than the speed of light.

Albert Einstein answered this question in **1905** with his theory of special relativity. Basically, it would take an almost infinite amount of energy to get us traveling even close to the speed of light. Scientists have gotten atomic particles to travel close to these speeds, but it requires so much energy. The amount of energy it would require to power a spaceship going this fast is far beyond our capabilities

SPEED

QUIZ TIME

1.Spaceships have carried:

a) People

b) Cargo

c) Animals

d) All of the above

2.Which two countries competed during the Space Age?

a) USA and USSR

b) England and France

c) USA and Japan

d) China and Brazil

3.True or False The United States was the first country to put a man in space.

a) True

b) False

In what year did America put a man on the moon?

a) 1950
b) 1959
c) 1969
d) 1982

Which spaceship is owned by Amazon's CEO Jeff Bezos?

a) New Shepard
b) Dragon
c) SpaceShipTwo
d) Orion

Which of Newton's Laws of Motion is used to help spaceships take off?

a) First
b) Second
c) Third
d) None

List the phases of a space mission in order from first to last: orbiting, re-entry, launching, landing

1. ________________________________

2. ________________________________

3. ________________________________

Author's Bio

The author is an educator and a passionate lover of children. He received his Bachelor of Arts in Biology from Edgewood College in Madison. Wisconsin and a second B.A. in Healthcare Management from Southern Illinois University Carbondale where he also completed his M.A in Engineering. He has a wealth of valuable experience dealing with children of varying age ranges. The author does not only love children, but he understands that when they have the best form of education, it has a far-reaching effect on the wider society. As a result, he is on a mission to make sure that children learn as much as they can.

References

1. A 21st century space capsule. Boeing. (n.d.). Retrieved November 6, 2021, from https://www.boeing.com/space/starliner/.

2. All shapes and sizes. ESA. (2011, October 12). Retrieved November 6, 2021, from https://www.esa.int/kids/en/learn/Technology/Spacecraft/All_shapes_and_sizes.

3. Bonsor, K. (2020, June 22). How self-healing spacecraft will work. HowStuffWorks Science. Retrieved November 6, 2021, from https://science.howstuffworks.com/self-healing-spacecraft.htm.

4. A brief history of space exploration: The aerospace corporation. Aerospace Corporation. (2018, June 1). Retrieved November 6, 2021, from https://aerospace.org/article/brief-history-space-exploration.

5. Demoulin, J. (n.d.). Rotational hand controller. NASA. Retrieved November 6, 2021, from https://science.ksc.nasa.gov/shuttle/technology/sts-newsref/sts-rhc.html.

6. Dragon. SpaceX. (n.d.). Retrieved November 6, 2021, from https://www.spacex.com/vehicles/dragon/.

7. Dunbar, B. (n.d.). Closing the Hatch. NASA. Retrieved November 6, 2021, from https://www.nasa.gov/mission_pages/shuttle/flyout/closeout.html.

8. Dunbar, B. (n.d.). The External Tank. NASA. Retrieved November 6, 2021, from https://www.nasa.gov/returntoflight/system/system_ET.html.

9. Dunbar, B. (n.d.). Shuttle Basics. NASA. Retrieved November 6, 2021, from https://www.nasa.gov/returntoflight/system/system_STS.html.

10. Garcia, M. (2015, April 12). Orion Overview. NASA. Retrieved November 6, 2021, from https://www.nasa.gov/exploration/systems/orion/about/index.html.

11. Howell, E. (2019, March 5). SpaceShipTwo: A flight path to space tourism. Space.com. Retrieved November 6, 2021, from https://www.space.com/19021-spaceshiptwo.html.

12. Jones, C. (2021, November 1). Often asked: How fast can spaceships travel? Family law. Retrieved November 6, 2021, from https://www.virginialeenlaw.com/help/often-asked-how-fast-can-spaceships-travel.html.

13. Kid's Guide to Space Shuttles and their parts. PartSelect. (n.d.). Retrieved November 6, 2021, from https://www.partselect.com/JustForFun/Guide-To-Space-Shuttles-And-Their-Parts.aspx.

14.McGrath, J. (n.d.). How do spacecraft re-enter the earth? | howstuffworks. How Stuff Works. Retrieved November 6, 2021, from https://science.howstuffworks.com/spacecraft-reentry.htm.

15. Mosher, D. (2019, July 19). Astronauts explain why nobody has visited the Moon in more than 45 years - and the reasons are depressing. Business Insider. Retrieved November 6, 2021, from https://www.businessinsider.com/moon-missions-why-astronauts-have-not-returned-2018-7.

16.Nagy, A. (2016, April 30). Here's how spacecraft dashboards evolved, and where they're headed. Gizmodo. Retrieved November 6, 2021, from https://gizmodo.com/heres-how-spacecraft-dashboards-evolved-and-where-they-1773640540.

17. NASA. (n.d.). Basics of space flight - solar system exploration: NASA science. NASA. Retrieved November 6, 2021, from https://solarsystem.nasa.gov/basics/chapter13-1/.

18. NASA. (n.d.). Guidance system. NASA. Retrieved November 6, 2021, from https://www.grc.nasa.gov/www/k-12/rocket/guidance.html.

19.National Geographic Society. (2019, December 10). The history of space exploration. National Geographic Society. Retrieved November 6, 2021, from https://www.nationalgeographic.org/article/history-space-exploration/.

20. Orion. Orion | Aerojet Rocketdyne. (n.d.). Retrieved November 6, 2021, from https://www.rocket.com/space/human-exploration/orion.

21. Perles, T. (2020, August 12). Vanadium: A green metal critical to aerospace and Clean Energy. Tech Briefs. Retrieved November 6, 2021, from https://www.aerodefensetech.com/component/content/article/adt/features/articles/37376.

22. Petruzzello, M. (n.d.). Will light-speed space travel ever be possible? Encyclopædia Britannica. Retrieved November 6, 2021, from https://www.britannica.com/story/will-light-speed-space-travel-ever-be-possible.

23.Redd, N. T. (2020, August 12). Meet 'Tenacity': 1st Dream Chaser Space Plane gets a name. Space.com. Retrieved November 6, 2021, from https://www.space.com/dream-chaser-space-plane-tenacity-tour.html.

24. Ryan, Jackson. (2021, May 3). NASA solar probe becomes fastest object ever built as it 'touches the Sun'. CNET. Retrieved November 6, 2021, from https://www.cnet.com/home/energy-and-utilities/nasa-solar-probe-becomes-fastest-object-ever-built-as-it-touches-the-sun/.

25. ScienceDaily. (n.d.). Spacecraft Propulsion. ScienceDaily. Retrieved November 6, 2021, from https://www.sciencedaily.com/terms/spacecraft_propulsion.htm.

26.Smith, B. K. (2006, February 13). What kind of fuel do rockets use and how does it give them enough power to get into space? Scientific American. Retrieved November 6, 2021, from https://www.scientificamerican.com/article/what-kind-of-fuel-do-rock/.

27. Smithsonian. (n.d.). Welcome to how things fly. | How Things Fly. Retrieved November 6, 2021, from https://howthingsfly.si.edu/ask-an-explainer/what-metals-are-used-spaceships-and-why.

28. Sparrow, G. (2021, June 30). How rockets work: A complete guide. Space.com. Retrieved November 6, 2021, from https://www.space.com/how-rockets-work.

29. The use of titanium on the spacecraft was effective in reducing the load. Space Daily. (n.d.). Retrieved November 6, 2021, from https://www.spacedaily.com/reports/The_use_of_Titanium_on_the_spacecraft_was_effective_in_reducing_the_load_999.html.

30. Walker, R. (2021, June 29). Why is it so difficult for a returning spacecraft to re-enter our atmosphere? Forbes. Retrieved November 6, 2021, from https://www.forbes.com/sites/quora/2017/02/03/why-is-it-so-difficult-for-a-returning-spacecraft-to-re-enter-our-atmosphere/?sh=6be0d1721177.

31. Wall, M. (2020, January 24). Four new U.S. spaceships may start launching people into space this year. Space.com. Retrieved November 6, 2021, from https://www.space.com/new-private-american-crewed-spaceships-2020.html.

32. Wall, M. (2013, September 12). It's official! voyager 1 spacecraft has left Solar System. Space.com. Retrieved November 6, 2021, from https://www.space.com/22729-voyager-1-spacecraft-interstellar-space.html.

33. What generates all the heat during re-entry when the space shuttle returns to Earth? Union University. (2003, March). Retrieved November 6, 2021, from https://www.uu.edu/dept/physics/scienceguys/2003Mar.cfm.

34. Williams, M. (2019, October 24). What is the space age? Universe Today. Retrieved November 6, 2021, from https://www.universetoday.

35. Wise, J. (2017, November 14). The 7 ships of the New Space age. Popular Mechanics. Retrieved November 6, 2021, from https://www.popularmechanics.com/space/rockets/g1388/the-7-ships-of-the-new-space-age/.

36. Zelazko, A. (n.d.). How many people have been to the Moon? Encyclopædia Britannica. Retrieved November 6, 2021, from https://www.britannica.com/story/how-many-people-have-been-to-the-moon.

37. Figures 1 – 30 Provided courtesy of Shutterstock. https://www.shutterstock.com/

www.ingramcontent.com/pod-product-compliance
Lightning Source LLC
LaVergne TN
LVHW070523061126
851857LV00023B/2132